W9-AMW-773

THE ALASKA PIPELINE

THE ALASKA PIPELINE

by Virginia O. Shumaker

Illustrated with
photographs and maps

Julian Messner

New York

Copyright © 1979 by Virginia O. Shumaker

All rights reserved including the right of
reproduction in whole or in part in any form.
Published by Julian Messner, a Simon & Schuster
Division of Gulf & Western Corporation, Simon &
Schuster Building, 1230 Avenue of the Americas,
New York, N.Y. 10020.

JULIAN MESSNER and colophon are trademarks of
Simon & Schuster, registered in the U.S. Patent
and Trademark Office.

Manufactured in the United States of America

Design by Philip Jaget

Second Printing, 1980

Library of Congress Cataloging in Publication Data

Shumaker, Virginia O
  The Alaska pipeline.

  Includes index.
  SUMMARY: Discusses the building of the Alaska pipeline, focusing on the
problems involved and how they were overcome.
  1. Alaska pipeline—Juvenile literature. [1. Alaska
pipeline. 2. Pipelines. 3. Oil] I. Title.
HD9580.U6A46   388.5′09798   79-13696
ISBN 0-671-32975-8

for Willem

ACKNOWLEDGMENTS

I wish to express my thanks to Sam Akin of the Alyeska Pipeline Service Company, Bob Olendorff, Bureau of Land Management and Tony Ciarochi, the U. S. Fish and Wildlife Service, all of Anchorage, Alaska; Lee Hoffman, my editor at Julian Messner; my parents, Herbert and Alice Olson of Richfield, Minnesota; and Robert and Dianne Hennessy King, who will always be Alaskans.

PHOTO AND MAP CREDITS

CONTENTS

Chapter 1

INTRODUCTION

Glorious it is to see
The caribou flocking down from the forests
And beginning
Their wandering to the north.
Timidly they watch
For the pitfalls of man.
Glorious it is to see
The great herds from the forests
Spreading out over the plains of white.
Glorious to see.

—an Eskimo poem

The animals stand quietly. Behind them, steep snowy mountains rise. This is northern Alaska. It is a quiet place. Towns are hundreds of miles away. A few small Eskimo and Indian villages dot the land. And animals live here—caribou, deer, wolves, mountain sheep, geese, grizzlies, and more.

During winter, animals find shelter from freezing winds in the mountain valleys. During the short summer, huge herds of caribou move down from the mountains to the flat lands to feed. The land slopes down for over a hundred miles, and ends at the edge of the icy Arctic Ocean.

This area is called the North Slope. For thousands of years, few people lived here. Even fewer came from the outside to visit. It was untouched wilderness. But something happened to change all that.

Chapter 2

DISCOVERY

Date: January, 1968
Place: North Slope, Alaska

Alaska Strikes It Rich!
Oil Is Discovered!

These are headlines from newspapers and magazines. Oil had been discovered near the Arctic Ocean in a place called Prudhoe Bay. Two oil companies—Atlantic Richfield and Exxon—made the discovery. Billions of barrels of oil lay far beneath the ground. The companies said it was the largest oil find in North America.

People knew that oil lay somewhere beneath this freezing land. It had come up to the surface in some places and made small pools. Eskimos had been using oil from these pools as fuel for years.

Exploration for oil had been going on in this area for over 40 years. Many wells had been drilled, but they had all turned out to be "dry," or without oil. Now, however, a rich oil strike had been made.

Oil that comes right from the ground is called crude oil. It is pumped up from deep underground pools or reservoirs. Although crude oil is of little use to people, many products are made from it. Among them are gasoline to run cars and oil to heat buildings.

Crude oil must go through changes, or be refined, to be made into products. But no oil refineries lay anywhere near the Alaskan oil field. The closest refineries were over 1,700 miles away in Washington and California.

How could this newly discovered oil be moved out of the North Slope? No highways or railroads went in or out of this wild place. Small airplanes could fly in. But billions of barrels of oil cannot be moved by airplane. It would be much, much too slow, expensive, and dangerous.

What about by ship? Prudhoe Bay lies just off the Arctic Ocean. Huge oil tankers carry thousands of barrels of oil to other parts of the world. But arctic waters are frozen into thick, solid ice most of the year. During summer, the ice melts a little and moves away from the shore for a mile or so. Could a tanker sail through? No one knew.

In other parts of the world, crude oil is often moved through pipes. Oil pipelines cross the deserts in Arab countries. Pipelines also carry oil from Texas, Louisiana, and Oklahoma to oil refineries in other parts of the United States.

Several months after discovering oil in Alaska, the oil companies announced they would try to build a pipeline out of Prudhoe Bay. It would be the first pipeline ever built in the Arctic.

The pipeline would move the oil south through Alaska. It would be about 800 miles long, and would cross three mountain ranges and hundreds of rivers and streams.

The town of Valdez (vahl-DEEZ) on Prince William Sound would be the pipeline's ending place. Ocean waters do not freeze

Among the items in front of this house, partly made from oil, are soap, toothpaste, phonograph records, plastic bags, wax paper, paint, luggage, footballs, carpeting, lipstick, and so on. You use many more oil-based products than you may think.

13

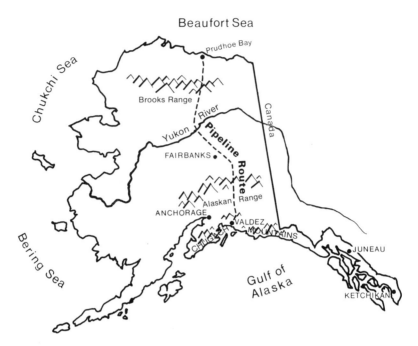

The pipeline route would cross three mountain chains: Brooks Range, Alaska Range and Chugach (CHEW-gatch) Mountains.

around Valdez and southern Alaska. Oil tankers could then carry the oil south to be refined.

Much of the land that the pipeline would cross was controlled by the federal government. The oil companies needed permission to cross these lands, and they were sure this would be easy to get. They were so sure that they began to bring pipe into Alaska right away.

But the oil companies were wrong. The permit to start the pipeline was not granted until 1974. For over five years, thousands of pieces of empty pipe lay stacked in Alaska. The reasons for this long delay are part of the story of the Alaskan pipeline.

Chapter 3

THE LONG DELAY

The Alaskan Native Claims

Who owned the land over which the pipeline would cross? For thousands of years, three groups of people have lived in Alaska—the Eskimos, Indians, and Aleuts. Like the Indian tribes in the rest of the United States, these people had lived in Alaska long before any white settlers moved in. They are called Alaskan natives, which means they were the first people to live there.

The Alaskan natives knew the land well. They had to. Their lives depended upon knowing the best hunting grounds and fishing areas. The land was used by and belonged to all of them.

But times changed. White settlers moved into Alaska. Towns and cities began to grow. Then Alaska became a state in 1959. The first governor and members of Congress were elected. The new government began to choose huge areas of land for state use. Among those areas were lands that were used by the native tribes.

The native people objected to the taking of their lands. They remembered what had happened to American Indians over a hundred years ago. These people had been treated unfairly when white settlers moved west. The settlers had taken Indian lands and had not paid for it.

The Alaskan natives did not want to be treated in the same way. They wanted to be well paid for their lands. And they wanted to keep some of the land for themselves.

The different native tribes joined together into one large group called the Alaskan Federation of Natives (A.F.N.). The A.F.N. told the United States government what they wanted for payment. The government leaders said they needed time to think about the demands. Meanwhile, the government placed a "freeze" on the lands. This meant nothing could be done with the land until the Alaskan natives' claims were decided upon.

Alaskan natives march in Anchorage, Alaska, to protest the unfair taking of their native lands.

What were the oil companies to do? They needed to build the Alaskan pipeline over these lands. But they would have to wait until the land "freeze" was removed before pipeline construction could begin.

The Alaskan native claims were settled in 1971. The various groups were given 40 million acres of land and paid almost one billion dollars. They were also promised jobs to help build the pipeline.

But during the delay to settle these claims, many people began to speak out against the pipeline. They were worried about the environment. Alaska was still an uncrowded land with clean air and water. How would the pipeline affect the arctic wilderness?

The Arctic

You may have heard the phrase "north of the Yukon." It means north of the Yukon River that runs through part of Canada and across Alaska. The Yukon is the third longest river in the United States, but no bridge had been built across it. The land beyond the river was always wilderness. No road had ever been built this far north.

This area north of the Yukon River is the last major wilderness area in the United States. Not one of the last—the last.

The North Slope is part of this wilderness. Temperatures drop to −50 or −60 degrees, making the arctic winters long and cold. But here is a fact that you might find hard to believe: this kind of land is one of the most fragile, or easily destroyed, on earth.

How can such a harsh land be so fragile? One reason is the small amount of sunlight it receives. The North Slope is near the "top" of the earth. During winter, the sun does not rise for several weeks. Days are gray, like evening. By early afternoon, the sky is black.

Almost one-half of Alaska's total land area is under control of the U.S. government. This land is used for parks, wildlife areas, forests, and military and oil reserves.

The sun returns during summer. For several weeks, days lengthen until they are 24 hours long. Temperatures climb, and some wild flowers bloom. But summer is short. And without much sunshine during the year, the plants grow very, very slowly. No forests grow. Trees may only be as tall as your ankles.

Nearly all the ground is covered with slow-growing grasses and mosses. This plant covering is called the tundra. Under the

Tundra covers much of the northern part of Alaska. It provides food for many kinds of arctic wildlife. If you look very closely, you can see caribou feeding.

tundra lies permafrost, which is ice, soil, and rocks all mixed together. This permafrost is frozen all year.

The tundra protects the permafrost from melting. When permafrost melts, it turns into mushy ice. And the mushy ice patch gets larger. It melts along its edges, destroying tundra as it spreads.

Some people felt this fragile land would be in danger if the pipeline were built. They even had proof. Many years earlier, government teams had been looking for oil on the North Slope. Bulldozer tracks could still be seen after 25 years. Garbage and equipment had been left behind. In warmer lands, garbage may slowly rot in the rain and sun. In the Arctic, garbage is frozen much of the year. Ugly piles of litter could last practically forever.

These oil drums were left on the arctic slopes by an early oil exploration team.

The Environment

Building the pipeline would surely change part of the land. But how much? Were there things that could be done to lessen the harm?

To help answer questions like these, Congress passed the National Environmental Protection Act (N.E.P.A.) in 1970. This act says that any work the United States government plans to do in an area must be carefully studied first. We need to find out how the work will affect the ecology and environment of the area.

Before the Alaskan pipeline could be built, many problems needed to be studied. Here are some of them:

- What would happen to the tundra? The oil companies first planned to bury the pipeline. A ditch, 789 miles long, would have to be dug. Trucks and other equipment would be driven back and forth over the fragile land. How could the tundra be protected?

- A buried pipe would run through permafrost. Crude oil comes from deep underground. It is hot—over 140 degrees. Hot oil would melt the permafrost. The pipe might sink into the mush and break.

- What about the wild animals? Would they stay in the area once the pipeline was built? Some kinds of wildlife need vast areas of wilderness in which to live. Caribou herds travel 3,000 miles in search of food. Some animals may become extinct without wilderness. Might this begin to happen in the North Slope?

- The pipeline route passes over three earthquake areas. If an earthquake occurred, the pipeline might break. Oil would spill over the land. It might go into rivers, killing fish and destroying parts of the rivers for centuries.

Large areas of wilderness are needed to feed wild animals. One Alaskan bear needs 100 square miles just to find food for itself and its young.

The worst earthquake every recorded in North America hit Alaska in 1964. This photo shows a street in Anchorage after the quake struck. Valdez, the site of the pipeline terminal, was badly destroyed in the earthquake. It had to be almost completely rebuilt.

- Ocean waters around Alaska are a big feeding area for ocean animals. Oil spills from ships had occurred in many other parts of the world. Might Alaskan ocean waters now become polluted from oil spills, too?
- Should this fragile land be opened to tourists? Engineers planned to build a road alongside the pipeline. When the pipeline was finished, people could drive into the wilderness. Would they litter as they do in the rest of the country? One person was afraid the road would be "800 more miles of ditches for people to throw cans in."

Debate about the pipeline became big news. Newspapers and magazines ran stories about it. Television shows told how different people felt.

Who wanted the pipeline built? The oil companies, of course. Alaskan oil would mean billions of dollars of profit for them. In 1970, eight companies joined together for the Alaskan pipeline project. They formed the Alyeska Pipeline Service Company. Alyeska means "the great land" in the language of the Aleuts.

Many Alaskans wanted the pipeline, too. Oil companies would pay the state for the use of state lands. Alaska would have money to build schools, roads, hospitals, and more.

Many American workers wanted the pipeline. They knew it would supply jobs. People would be needed to build the pipeline and equipment for it, oil tankers, and so on.

Some people felt we should start building the pipeline right away because the United States was going through an "energy crisis." Fuel was in short supply. Cars had to wait in long lines at gas stations.

Other people felt differently. They said that Americans were wasting too much fuel, and instead of drilling for *more* oil, we should work hard to use less.

24

A shortage of heating fuel meant no heat in this Kansas classroom in December, 1973. Temperatures in the unheated room dropped to 42 degrees.

One environmentalist said, "The oil has been there for millions of years. It will keep. We should take time to find a way to get the oil without wrecking the land and its life."

Other Ideas

Was the Alaskan pipeline the best way to get oil out of the North Slope? People wondered if oil tankers could sail 4,500 miles from the east coast of the United States to Prudhoe Bay and back again. Then a pipeline would not be needed.

Humble Oil Company sent out a huge ship, the *SS Manhattan*, to try. The trip was completed. But the company decided that carrying oil out by ship would be too difficult.

Another idea was to run the pipeline through Canada to the midwestern United States. The oil was needed more in that part of the country. Although a pipeline across Canada would be longer, it wouldn't pass over earthquake areas. There would be no danger of ocean oil spills.

And there was another important reason to think about an oil pipeline across Canada. Whenever oil is found in the ground, natural gas is found with it. They are made in nature at the same time. The huge reserve of natural gas found in Prudhoe Bay would be sold as fuel. The safest and cheapest way to move the gas would be by pipeline to the midwest.

Talks between Canada and the United States were going on to make plans for the gas pipeline. Why not leave the Alaskan oil in the ground until the gas pipeline was built? Then one route could be used for two pipelines. Many people felt this was a good idea.

But the oil companies were against it. They had already spent a great deal of money planning an Alaskan pipeline. In the end, the idea for a trans-Canada pipeline was defeated.

Chapter 4

CHANGES IN THE PIPELINE

The debate continued for several years. During this time, many people worked hard to stop construction. Environmental groups said that building the pipeline would be breaking the National Environmental Protection Act. They felt that not enough tests had been done to make sure the pipeline would be safe.

From 1970 to 1973, the oil companies and the United States government made thousands and thousands of tests to build a pipeline that would protect the land. Such a pipeline had never been laid in the Arctic before. Scientists and engineers were faced with special problems for the first time.

Important changes in the pipeline plans were made. Here are some of them:

- Scientists tested soil and permafrost samples along the pipeline route. The tests showed that heat from buried pipeline would melt the permafrost in some places. The pipe might sink into the ground and burst. So instead of burying the pipeline, over one-half of it would be laid above-ground.

A geologist, or earth scientist, examines part of the Alaskan pipeline route. Scientists drilled holes along the 789-mile route to test soil and permafrost.

- The pipeline would cross about 350 rivers. Scientists studied fish living in the rivers. They learned when the fish lay eggs and when they migrate to new waters. Plans were made so the construction would not block the rivers during these times.

- How would animals pass by the aboveground pipe? Short test pipes were put in place. Moose, bear, and other animals walked under the pipe or passed over on ramps. But huge herds of caribou would be crossing the pipeline route at several spots. They could not pass by the pipe so easily. Environmentalists said the pipeline had to be buried in certain spots to let the caribou by. But fragile permafrost was under the ground at these spots. So designers made plans to cool or refrigerate parts of the buried pipe.

- Hundreds of miles of tundra would have to be destroyed while building the pipeline. Scientists studied seeds and plants from arctic lands in Russia, Finland and Canada.

This scientist collected a sample from Alaskan waters to help test for oil spills.

They found fast-growing plants to protect the land until the tundra grows back. They would plant these seeds as soon as pipeline work was completed.

A gravel work pad would be laid all along the pipeline route. All work would take place on this pad. This would protect the tundra, too.

- Special posts were designed to hold the pipeline as it passes over earthquake areas. The posts would allow the pipeline to move 20 feet back and forth or 10 feet up and down without breaking. These plans should work if a future earthquake is the same size as earlier ones. If a stronger earthquake strikes, no one knows what might happen. The pipeline may be safe. Or it might break.

- Special gates, or valves, were designed to be placed along the pipe at several points. When open, these valves let the oil pass through. But if a break occurs, the valves would shut. Oil would stop flowing so less would spill out.

All these and even more safety ideas were put into the pipeline designs. Some problems were solved. But it will be years before we know if all the solutions will work.

On November 16, 1973, the Trans Alaska Pipeline Authorization Act became a law. It meant that Alyeska could start construction.

This valve can stop oil flow if a leak occurs further down the pipeline.

Chapter 5

THE FIRST STEPS

What exactly needed to be built? The pipeline, of course, would be laid along 789 miles. To keep oil moving all that way, nine pump stations had to be built. Three more could be added later if needed. The huge pumps would push the oil up over mountains and along flat lands.

The pipeline would end at the Valdez terminal. Eighteen giant storage tanks were needed to store the oil. Each of these tanks would be like an empty can as big as a building.

Docks had to be built out into the water for oil tankers to stop and fill up with oil.

Equipment was needed to keep the air and water clean around the terminal.

And the communications system that was to check on the oil as it flowed from Prudhoe Bay had to be built.

Where to begin? If you have ever seen a highway being built, you know that some workers build bridges while others pave the road or clear the route of trees farther on ahead. Work goes on at many places at once. The same thing was planned for the Alaskan pipeline.

Construction Camps

Thirty-one sites spaced along the route would be construction camps. Buildings to house workers and store machinery were erected at each camp.

The rush to move equipment to these camps began right away. But how could trucks cross the Yukon River to reach sites in the

A line of trucks and tractors moves past icy mountains, heading north of the Yukon River. During early months of construction, up to one hundred trucks and tractors crossed the river every day.

far north? A bridge would not be ready for at least a year. Freezing winter weather helped solve the problem. Trucks and tractors could drive across the river's frozen surface.

To make sure the ice would hold the heavy loads, an ice bridge was made. Water was pumped out from under the surface and sprayed over the river where it froze. Layer upon layer of ice froze to make an ice bridge six feet thick. But time was short because the ice would melt with the spring thaw.

A race with winter began. Giant bulldozers crossed the Yukon and scraped a smooth path over the ice and snow. This "ice highway" went north toward Prudhoe Bay.

Trucks and tractors then roared back and forth across the river and along the ice highway 24 hours a day. Drivers went through mountains and along steep cliffs to reach camp sites. They unloaded equipment and headed south again for another load.

One truck driver said, "I wouldn't go through that winter again for a million dollars. But I wouldn't give up the memories for ten million. We helped each other out. If you met another driver in trouble, you helped out. If you didn't, he might die."

People hired by Alyeska who worked north of the Yukon had to take special classes to learn about working in cold weather. They found out what kind of clothing to wear and what emergency supplies to carry with them. People going out on their own, such as a truck driver, were to carry a sleeping bag, a medical kit, emergency food for five days, a knife, flashlight, map, matches, and plenty of warm clothing.

Airplanes were also needed to speed camp materials north, so landing strips were cleared. Huge cargo planes flew in lumber and fuel. They also brought in small mobile camp buildings that had been made to fit within the plane's cargo space.

The construction camps began to take shape. The mobile buildings, which looked like trailers or mobile homes from the outside, were placed in rows.

Inside most of the buildings were rooms for workers, with beds, dressers, curtains, and carpets already in them. Other buildings contained huge kitchens where meals would be cooked. Supplies like soap, towels, pots, and pans had to be flown in. Plans were made for buildings to contain a library or a heated swimming pool that workers could use during their time off.

The Hercules cargo plane was used to fly equipment and camp buildings to construction sites. Buses like these carried workers from camps to the pipeline.

DISCARD

Carpenters built walkways between the buildings so workers would be able to move around much of the camp without going outside.

Finding a Pipeline Job

In the meantime, people began pouring into Alaska looking for high-paying pipeline jobs. The pipe welders and oil drillers who were hired had worked on pipelines before. But the Alaskan pipeline would be different.

One worker said, "It's a job I've never seen anywhere in my life before. I've been in deserts. I've worked in many countries. But I've never worked with so much snow in my life."

Huge cranes were needed at construction camps to help move camp buildings into place.

Men and women were hired to work in construction camps as dishwashers, cooks, laundry workers, nurses, librarians, carpenters, and more.

The Alaskan pipeline was one of the first big construction jobs where women were hired to do tough, outdoor work. Some of them drove trucks, directed traffic, ran heavy equipment, or helped install pipeline posts. Others had more traditional jobs such as secretaries, cooks, and housekeepers.

Not all the women worked on construction. This woman painted serial numbers on pipes at Prudhoe Bay to help Alyeska keep track of the more than 101,000 sections used in the pipeline.

This Alaskan woman went to a special school to learn how to operate a crane. The top student in her class, she is shown here operating a crane at Pump Station 3.

Pipeline jobs were filled fast, but people kept coming into Alaska, hoping to find work. Cities like Fairbanks and Valdez grew more and more crowded. Many Alaskans were unhappy over all the newcomers, especially when food and housing costs rose, and crime and pollution increased. The state government put ads in newspapers telling people *not* to come.

Some people waited months for work; others gave up and left. But thousands found jobs.

The Haul Road

As workers were hired, they were flown into construction camps to start work. The first task was to build a gravel road along the 789-mile route. This road, called the haul road, would carry heavy trucks and machinery up and down the route. The road would also help protect the land underneath it.

Road construction began in April, 1974. Bulldozers pushed tons of gravel out of dry riverbeds and rocky hills. Trucks carried the gravel back to the road site, where it was dumped and spread onto the surface.

Summer came and days grew longer and longer. For weeks, work on the haul road went on in sunlight for 24 hours a day. The haul road was finished in September.

During this time, environmentalists were busy along the route, too. They found rare birds living near the route of the haul road, and said it had to be moved to protect the birds. Alyeska moved the road five miles away from the nesting area.

Litter and sewage from the camps polluted some areas. Soil and rocks were often pushed accidentally into streams, which dirtied the water and could harm fish.

Environmentalists often stopped construction until a problem was corrected, but sometimes they discovered a problem too late. At other times, the pipeline builders did not want to do what the environmentalists said. Any delay cost Alyeska money. But without environmentalists, many more mistakes would have been made and left uncorrected.

This photograph was taken just before midnight in June, 1974. During summer, the sun is in the Arctic sky 24 hours a day for several weeks. These bulldozers are working on the haul road.

Chapter 6

THE PIPE IS LAID

Preparations for laying pipe took nearly one and a half years. But once the pipe laying began, it usually went along smoothly.

The pipe itself is made of a strong metal and looks silvery and shiny on the outside. A person four feet tall could stand up inside the pipe with his or her head just touching the top.

To bury the pipe, powerful machines scooped out big bucketsful of earth. Pieces of pipe 80 feet long were lowered into the ditch. Sometimes the ditch curved as it went over the top of a hill. A special machine bent the pipe until it matched the curve of the land.

The pieces of pipe were welded together, and then special cameras took photographs of each weld. The photos showed if it was solid and strong. After welding, the pipe was wrapped in tape, lowered into the ditch, and buried.

To install pipe aboveground, machines drilled holes into the land 20 to 45 feet deep and about 60 feet apart. Workers sank pairs of strong metal posts to give support. Pipe was placed on top of the beam and welded together. Again, the weld was photographed and checked.

Pipeline construction continued through summer and winter. Some work went on indoors at pump stations or at the Valdez terminal.

The ditch dug for the pipeline was about 12 feet deep and 8 feet wide.

Welders used hot liquid metal to join together two pieces of steel pipeline. Welders often worked under tents for shelter from the weather.

This pipe was bent to follow the shape of the ditch. It is being lowered into the ditch where it will be welded to the pipe section at lower right.

All welds on the pipeline were required to be checked with an X-ray photo. Also, inspectors checked welds from the outside. They sometimes used mirrors to help them check hard-to-see spots.

A face mask made of a special material helped keep this worker warm.

Two-story buildings were erected as living quarters for construction workers at nine pump station sites along the pipeline route.

People could sometimes work outdoors for only thirty minutes before they had to warm up. Clothing would often include two sets of long underwear, special boots and overalls insulated to keep out cold, wool and leather mittens, and at least one face mask.

One worker walked 13 miles in the winter to help a stranded

bulldozer. He recalls, "I had three face masks on and it was just frozen solid. You could hit your face and it was like hitting a rock."

Machines were kept running 24 hours a day through the winter. If they were turned off, the engines froze and were very hard to start again.

For dinner, a worker could choose from soup, five kinds of salad, four cooked vegetables, lamb chops, roast beef, chicken, seven kinds of pie, cakes, ice cream, cookies, pudding, coffee, tea, milk, cocoa, fruit drink, or soda pop.

The construction camps were the only homes the workers had while they were on the pipeline. After working hard all day, they wanted to relax and be comfortable when they had time off.

Food—lots of it—was served in the camps 24 hours a day. People could listen to music, swim in the heated pool, read the newspapers that were flown in everyday, play ping pong or chess, or watch a movie.

Trees were planted inside the camp at Pump Station 1, because all the workers saw outside were snow and ice. The trees were nicknamed the Prudhoe Bay Forest.

Medicines were flown by helicopter to all parts of the pipeline. To keep the medicines from freezing, they were stored in special heat-holding chests.

The construction camps also supplied medical care. Working in the extreme cold can be dangerous, especially when using heavy construction machinery. Bare skin can freeze within seconds, and accidents happen more easily because workers are cold and stiff. People were flown out to hospitals if a serious accident occurred.

49

Winter weather was not the only problem. Some of the Alaskan pipeline crossed areas that were especially rugged and hard-to-reach. The Keystone Canyon was one of these spots. A river flows at the bottom of the canyon which has steep rock cliffs rising on both sides. The pipe had to be laid at the top of the cliffs. Workers and equipment had to be flown up by helicopter. Laying pipe on this four-mile stretch took ten months.

Thompson Pass was another steep place along the route. Here, welders tied themselves to the pipe as they worked on it. This helped stop them from slipping down the hill.

During 1975, pipeline construction ran into serious trouble with welding. Alyeska had agreed to check and double check each weld. A weak weld could burst and spill oil. Someone found that many welds had not been checked as required. Some weak welds had been buried before being repaired.

News about the faulty welds leaked out. There were stories in newspapers and on TV and radio. A Congressional committee held meetings to get more information. Almost 1,000 welds were found to be either unchecked or with weak spots. Alyeska corrected all the problem welds.

As pipeline construction went on, one job took place nearby that had nothing to do with oil or pipes or the environment. Archeologists, scientists who study lives of ancient people, wanted to learn more about the ancient Alaskan tribes. They were afraid the pipeline construction would destroy whatever clues to these people's lives that had been left behind.

A piece of machinery is being air lifted to the construction site atop the wall of Keystone Canyon. Smaller machines were lifted whole. Larger ones were lifted in pieces and put together again at the top.

The archeologists first decided where the ancient people would have probably traveled. Then they went to work just ahead of the pipeline crews looking for clues. The archeologists dug into the soil and studied each shovelful of dirt. They found many items that had been left behind by ancient tribes: fishing hooks, harpoons, arrowheads, and parts of sleds made from whalebones.

These archeologists are carefully digging through soil at a place where ancient people once lived.

Chapter 7

FINISHING UP

By mid-1977, the pipeline was finished and had been tested for strength. To do this, the pipes were filled with water. Then even more water was "squeezed" into the pipes, or put in under pressure. The water pushed against the sides of the pipe and burst through any leaks or weak spots, which could then be repaired.

All nine pump stations along the route were ready to start pumping.

In Valdez, the 18 huge storage tanks were built with high walls, or dikes, around every two tanks. If an earthquake or explosion ever breaks the tanks open, the dikes will stop the oil from flowing into the harbor.

Air and water cleaning equipment in Valdez was ready, too. The gases that rise from stored oil would be kept from polluting the air. Three ballast treatment tanks would help keep harbor water clean. Ballast is the water that empty oil tankers carry to add weight so they sail better. Ballast water becomes oily and dirty from the inside of the huge tanks. Often, this oily water is dumped into the ocean when a tanker is ready to take on oil. But in Valdez, ballast water will be cleaned before it is pumped back into the ocean.

This air view shows the Valdez terminal nearing completion. Two more tanker docks still need to be built.

Four huge docks were ready for the first tanker to berth and load up with oil.

A huge computer system that controls the movement of oil through the whole pipeline was ready.

The pipeline route had been cleaned up, and ground that had been dug up was being reseeded.

This man at a pump station can talk to the Valdez terminal over the telephone to report any oil information or problems. In front of him is a board giving up-to-date information on the flow of oil through the pipes.

Seeds, water, and mulch are sprayed over dug-up areas to start grass growing there.

Start-Up

On June 20, 1977, the oil began to flow from Prudhoe Bay. For almost 500 miles, everything went well. The oil moved slowly through the pipe at about a mile an hour.

Then the oil flowed into Pump Station 8. Suddenly, there was an explosion, and black oil sprayed out. The oil flow was

Oil from Prudhoe Bay enters the Alaskan pipeline at Pump Station 1 on its journey to the terminal at Valdez. The zigzag pipe allows for small movements in the pipeline caused by changes in temperature through the seasons.

stopped right away. One man was killed, and the pump station was badly damaged. What could have caused the explosion? Alyeska said a worker had accidently opened a valve too soon.

Ten days later, oil started to flow again. The first barrel of Prudhoe Bay oil reached Valdez on July 28. At the dock waited the oil tanker, *ARCO Juneau*. The ship was filled with 824,000 barrels of North Slope oil, and sailed south with its cargo.

The Alaskan pipeline was operating.

In case any oil spills while a tanker is being filled, a containment boom surrounds the ship. This will help keep the oil from spreading.

Chapter 8

WHAT'S AHEAD?

Since start-up, the Alaskan pipeline has had some problems. Oil has been spilled at pump stations and at Valdez. People have tried to damage the aboveground pipe. In February, 1978, an explosion made a small hole in the pipeline. About 8,000 barrels of oil gushed out over the land before the flow could be stopped. The person who set the explosion was arrested.

While oil is being pumped from the ground at Prudhoe Bay, natural gas is surfacing, too. Plans are still being made to build a pipeline across Canada to move this gas to the midwestern United States. Construction may begin soon. Many people still feel that the oil pipeline should have been built along the same trans-Canada route.

What about the future of the Alaskan pipeline? It will probably take about 20 years to pump out all the oil from the North Slope. No one knows what might happen during those years. Building the pipeline was not like building another bridge or another new skyscraper. Alaskan pipeline problems were new and needed new solutions.

Will an earthquake split it open? Will wildlife grow used to living near it? Will fish stay in Prince William Sound near Valdez, or will oil spills drive them away? No one knows for sure.

Pools of oil spread over the ground where an explosion damaged the pipeline. About 8,000 barrels of oil spilled before the leak could be stopped.

INDEX